S0-CFH-831

EARTHSCAPES

Landforms Sculpted by Water, Wind, and Ice

by Jerry Wermund

EARTHSCAPES
Second Printing 2007
Book Design by Tim Wermund
Text Copyright © 2003 by Jerry Wermund
Photographs Copyright © 2003 by Jerry Wermund except where credited to another scientist (page 48)
All rights reserved. No part of this book may be used or reproduced in any manner whatsoever without written permission except in the case of brief quotations embodied in critical articles and reviews.
Copyright Rockon Publishing 210 Hy Road Buda, TX 78610

ISBN 0 – 9726255 – 0- X

Printed in Hong Kong through Creative Printing USA

Acknowledgements

to my wife Susan who said try

to my mentor Kathi Appelt who said you can

to my critique peers Dianna Aston, Anne Bustard, Betty X. Davis, Meredith Davis, Jimmy Hendrix, Frances Hill, Lindsey Lane, Greg Leitich, Jane Peddicord, Cynthia Leitich Smith, and Andrea Warkentin who said improve.

to Amanda Masterson who gave a final edit.

to Charles Kreitler, Gary Kocurek, Louis J. Mayer, David Noe, Jay Raney, C. J. Waffington, Tim Wermund, Johnathan White, William White and Charles Woodruff Jr. who contributed photographs.

MOUNTAINS

Thrust upward by internal forces,
mountains pierce the sky,
a momentary achievement
to be lost to erosion,
carried away
to the seas.

JOY FOR A HIKER

A dome,
rounded hill,
former mountain
eroded to its core,
bare granite remains
cracking and peeling
like skin off an onion.

CASCADES' WAR AND PEACE

A volcano rumbles and thunders,
flashes fire and erupts in a storm,
explodes ash and smoke heavenwards,
spouts fiery rivers down its flanks.
Falls asleep again.
Lies inert into future centuries,
with an innocent beauty,
an upside down cone of black and green
capped by a snowfield of whipped cream,
framed by a cyan sky.

SATAN'S SPIRE

Silvery black
basaltic rock,
six-sided pillars
stacked together,
the frozen spine
of an ancient volcano,
a neck-breaking view!

LANDSLIDE

Instantly,
    out of an arcuate scar, a hillside soaked with water
      glissades, swoops, swooshes down to a level,
        leaving frowns across its face,
          ending in a hump like a raised big toe.

TALUS

Weathered
broken off
steep rocky cliffs,
masses of fragments,
chunks, slabs, chips, and slivers,
fall, bounce, and tumble, pulled downward
in gravity's power, slip and slide, into jumbles and piles,
forming blankets and sheets of debris along a trail,
down - down the slope burying trees, bushes, and grasses.

13

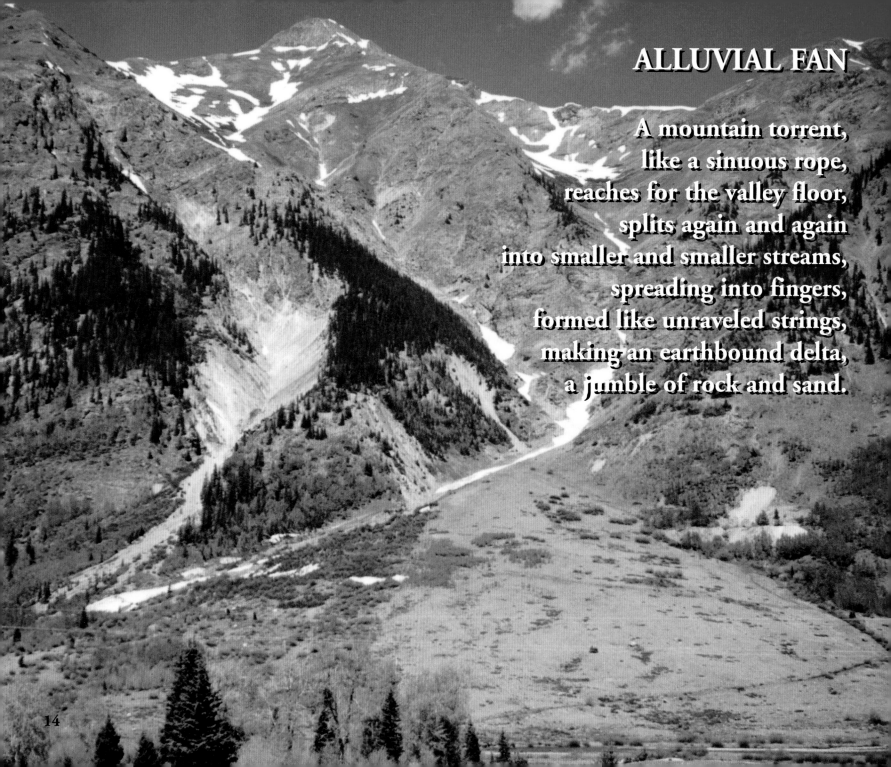

ALLUVIAL FAN

A mountain torrent,
like a sinuous rope,
reaches for the valley floor,
splits again and again
into smaller and smaller streams,
spreading into fingers,
formed like unraveled strings,
making an earthbound delta,
a jumble of rock and sand.

14

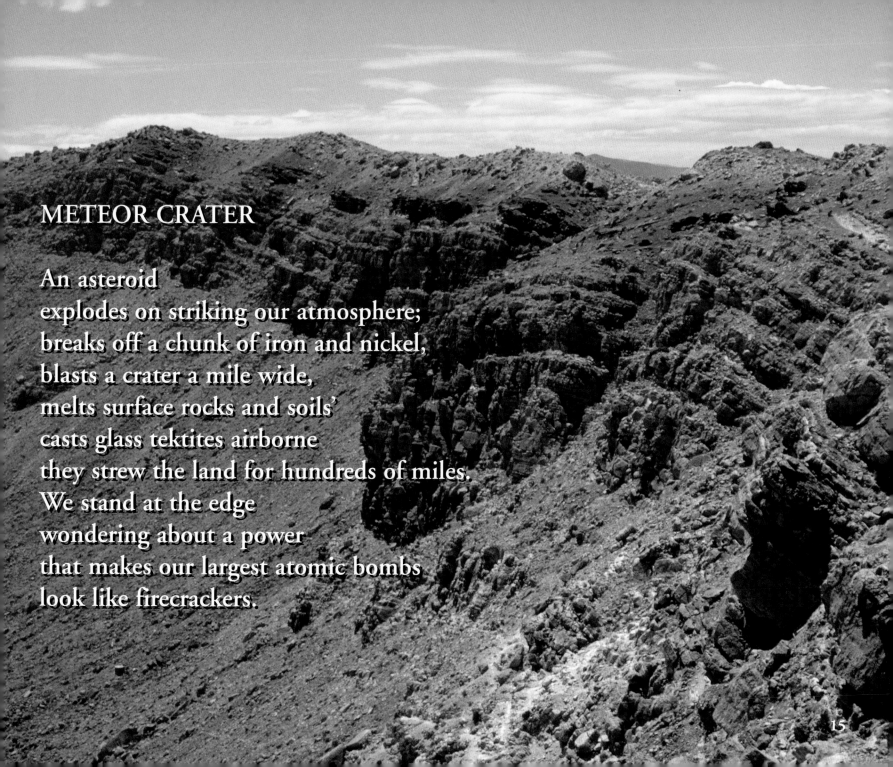

METEOR CRATER

An asteroid
explodes on striking our atmosphere;
breaks off a chunk of iron and nickel,
blasts a crater a mile wide,
melts surface rocks and soils'
casts glass tektites airborne
they strew the land for hundreds of miles.
We stand at the edge
wondering about a power
that makes our largest atomic bombs
look like firecrackers.

MEANDERS

Pushed by
forces
of a spinning Earth,
stream meanders wind
back and forth
and
back and forth
like an agile snake,
flowing down slope
to an end in
still water.

CUT BANK

The outside bank of a meander loop
stands precipitous and shadowy,
a weakened slumping alluvial pile,
a grave for trees and shrubs.

POINT BAR

Across from the Cut Bank
    a prong of sediment
        juts into the stream,
          a potpourri of
        sand and gravel –
     a library of
    upstream rocks
dropped when floods cease.

BRAIDED STREAM

Between valley walls of overburdened streams,
interlacing multiple ribbons
flow downstream
surrounding the
tapered ends of
teardrop
sand bars.

FARMER'S PLAGUE

The gully head,
a dragon's mouth,
devours his rich
topsoil and sediment
stealing life from pastures
and cropped fields,
passing its precious load
down a
destructive
channel.

SINKHOLES

In limestone country, underground rivers flow
slowly slowly slowly
silently silently silently
beneath our feet;
dissolve away roofs of caves,
collapsing them with a crash
louder than a car wreck; leaving pits
often circular; entrances for us
to explore new caves.

CANYONS

From the valley floor
in deep dark,
shear walls
climb upward
seeking
daylight.

Cliffs
press
inward
confining
bone-dry
channels
awaiting
flashy
floods,
perhaps
for centuries

BEACH

Where sea meets land,
kissing, caressing,
or battling to win shoreline;
it steals then returns sand
to a playground
for tots to build castles;
for collectors to find favorite shells.

COASTAL FOREDUNES

Offshore winds sweep off the beach,
caught and held in the first vegetation.
Sand climbs into dunes -three men high -
builds a barrier bracing against
relentless attempts of the sea
to capture the land.

TIDAL FLATS

Gentle sloped lowlands
encircle the bay.
Like breathing in and breathing out,
tides flood then bare the land.
At high tide, crabs, clams, and snails
rise to the surface from their burrows.
During low tides, predators
feast off the flat dining room table.

HOOK

Currents along the
shore convey a load of sand
down the beach front, seeking a headland,
where, dropping their burden,
a new sand deposit,
curved like a scimitar,
bars entrance
to a bay.

IN PRAISE OF MARSH

Spartina alterniflora
blankets miles and miles and miles
with brackish cordgrass meadows,
captures current borne sand and mud,
extends the land into bays,
filters pollutants seeping and streaming from the shore,
shelters juvenile creatures:
shrimps and crabs, turtles and toads,
fish and birds, hiding among stalks and rizome
roots of Spartina alterniflora.

NOTE: Spartina alterniflora is the most abundant marsh grass species in marine and brackish coastal waters of the United States. In many areas, we are now replanting this species by hand to recover lost marsh.

29

SEA STACKS

Where steep
headlands
of hard rock
abut the sea,
huge waves
attack and part
the jointed rock,
leaving lone towers
surrounded by the sea,
remembering they
were once land.

FJORD

Drown a glacial
valley flowing to the sea;
you get a fjord.

BARCHAN DUNES

A horizontal wind
unwinds from a dust devil,
rolls and bounces quartz sands
over a rippled surface,
up the gentle foreslope
onto a high crest.
There, sand grains fall,
tinkling unheard,
down the steep backslope.

Dunes spread their wings forward,
mimicking the quarter moon.
Grain over grain over grain,
barchans creep across the desert floor.
Sunset tints the foreslope,
casts a black shadow,
on the dunes' lee side.

HAMMADA

In mid-desert,
travelers drive
forever
over ground
flat as a floor;
no ups; no downs.
Its sands were blown away,
leaving behind
a pavement
of tiny pebbles.

AEOLIAN ARCH

How can it be?
Half a
lopsided,
sandstone
bagel
vaults over the desert floor.

SAHARA WADI

Living plants on the wadi floor
proclaim "Rain once fell –
could again – to fill
this dry wash with Life!"

On parapets of the fort,
ghosts of French Legionnaires;
and of black-robed Tauregs
float on simmering heat waves.

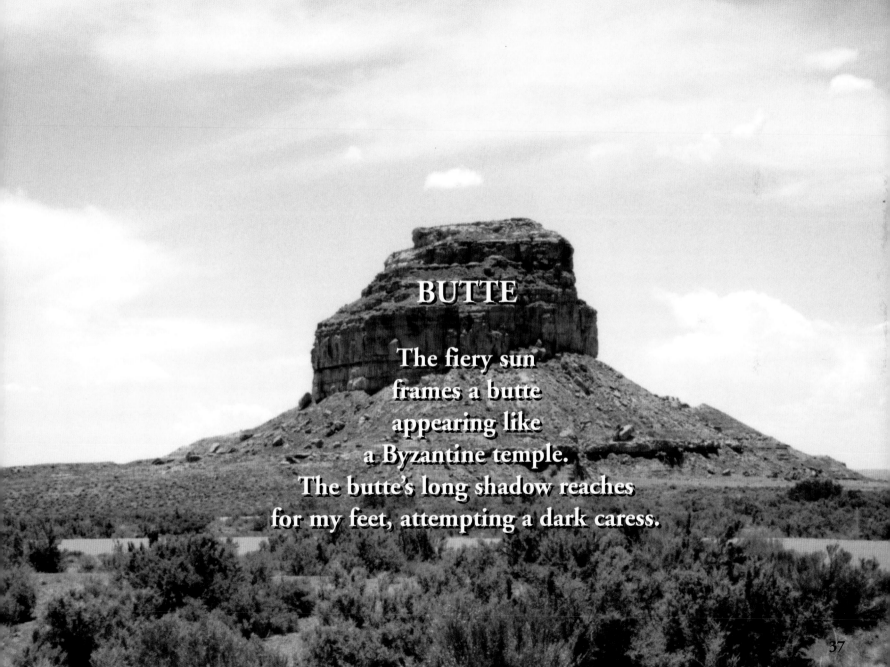

BUTTE

The fiery sun
frames a butte
appearing like
a Byzantine temple.
The butte's long shadow reaches
for my feet, attempting a dark caress.

PLAYA

A temporary lake –
a playa –
evaporates,
leaving a veneer of
alabaster, golden, rose, and emerald salts
on the reflective floor
of its solar oven.

MOUNTAIN GLACIER

Below a cirque
a thick tongue of ice
creeps down slope
bull-dozing soil and rocks,
trenching its own
deep U-shaped valley.
Cracks and crevasses
inscribe the ice surface
from valley wall
to valley wall.
with primitive writing.

CIRQUE

A mountain snowfield
freezes anew
winter upon winter.
Ice compacts and creeps
again and again,
plucking rock from the peak
over and over.
The glacier flows
down - down,
creating a bowl,
cupping an azure lake,
amphitheater for cloud dramas,
stage for a marmot choir
who whistle a wind borne chorus.

HANGING VALLEY

In the hanging valley,
gouged by
an ancient glacier,
an almost dry chute
remembers
its cataract past.

43

DRUMLINS

In evidence of his passing
through New York
through Wisconsin,
the glacial god for continents
set down his giant cigars
onto the land,
side by side,
row after row.
Buried beneath a
gigantic ice sheet.

HUMANSCAPES

Humans employ
machines and industrial materials
to overprint a rigid geometry –
lines and polygons
arcs and circles –
onto the land
where nature used
water, wind, and ice
to sculpt
its gentle freeform Earthscape.

Illustrations